The Healing Power of Awareness

How to Recover Faster and Less Painfully Using Sentient Awareness

The Healing Power of Awareness

How to Recover Faster
and Less Painfully
Using Sentient Awareness

Tom Richards

ISBN   978-0-578-10054-8

Published by:

Sentient Care Publications

1218 Roosevelt Avenue

Glenview, Illinois

60025 USA

www.sentientcare.com

First Edition 2012

First Printing

Contents

In Appreciation

With gratitude to Dr. Pierre Morin for awakening me to my incredible internal and external creative freedom,

to Drs. Arnold and Amy Mindell and their associates for developing the new practical paradigm Process Work,

to Cindy Trawinski and Stan Tomandl for their insights, guidance, editorial gifts, and friendship,

and to Marissa and Margaret for their courage to explore, and permission to publish their explorations.

Preface

This book is a treat to read. Tom is a gifted facilitator and teacher with a quickness in style and heart that provides people easy access to their dreaming natures in fun, creative, respectful ways.

Tom's work also has an "Ah Ha!" quality to it that can inspire people to just step into the unexplored a bit further and deeper. He demonstrates significantly enhanced healing and pain reduction for patients who explore sensory-grounded signals and follow nature using sentient awareness.

This approach, which suggests a new healthcare paradigm, can benefit professionals, caregivers, and family members who yearn for ways to access the sacred, but are not given enough time with their patients and loved ones, nor training, to even have a conversation about someone's life, let alone enter into the dreaming of a person's symptoms.

This book will be inspiring for people interested in body healing, including anyone yearning for more of themselves and the ones they are working with, to be recognized and interacted with. Here is a teaching tool as well as an inspiring story for all who read it.

I share Tom's dream of unleashing the incredible and mysterious healing processes inherent in the human body.

Ann Jacob
Founder, Coma Communication
Mentor, Sacred Art of Living Center

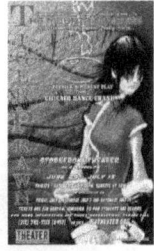

Introduction

This book demonstrates the healing power of sentient awareness when applied in concert with surgical procedures and rehabilitation, and suggests an intriguing new healthcare paradigm. The benefits to patients often include faster recovery, diminished pain, and reduced chance of relapse or recurrence.

Just as the true nature of electricity remains a mystery, we are in our infancy when it comes to understanding the incredible healing processes inherent in our bodies. And just as we can harness electricity despite its mysterious origin, we can enhance our bodies' natural healing processes with sentient awareness.

Sentient awareness explores the dreaming and meaning contained in sentient perceptions; those subtle, minute flickerings of body feelings, movements, sounds, and images that accompany body experiences and symptoms.

Finding the meaning contained in body symptoms, can relieve our body symptoms and pain, by easing internal emotional conflicts, and by engaging our "unconscious" systems, including our autonomic and immune systems. By providing our body with essential meaningful experiences about our life, sentient awareness unleashes inherent energies, and makes them available for life and living and healing.

Sentient awareness is an act of reverence that compliments and enhances the whole person; mind, body, and spirit.

Marissa (and her right ankle)

Photograph by Kristie Kahns

Marissa and Surgery

Marissa is a vibrant twenty-four year old professional dancer and dear family member. For several years she has danced through injuries to her right ankle resulting in; a loose bone fragment, chronic inflammation, and intense pain. She has exhausted noninvasive therapeutic treatments of all types. Because of chronic pain and swelling in her ankle Marissa is now considering surgery. Her surgeon, a specialist experienced in dance injuries, tells her: *Without surgery you can live with this condition, but you won't dance . . . and you would have to give up running.* Marissa courageously decides surgery is necessary for her career and well being.

Excited, nervous, and anxious

Marissa: *Before surgery I felt excited, nervous, and anxious. I felt excited because I felt I was going in a direction of finally fixing the problem, a problem that had been vexing me for years, before it was absolutely too late, before my professional dancing shelf-life had expired. It was an exciting thought that I would be able to dance again without constant hesitation and pain in my ankle.*

I felt nervous because I couldn't shake this feeling that there was something I hadn't tried. I didn't want to go under the knife. I couldn't necessarily afford extensive acupuncture therapies, but I felt nervous that maybe there were other methods of fixing the damage that I had overlooked.

I felt anxious because I couldn't really guess what improvements I was going to see. No one could guarantee any specific recovery. They could only tell me what the procedure would fix physically. I had no idea how long it would take MY body to recover. I had no idea how much pain would be involved, although I have to admit, this was somehow the least of my worries. No one could guarantee me a set time to return to the stage. It was nerve-racking. What if surgery DIDN'T fix the problem . . .

Wake me when it's over

However, after years of treating and caring for her ankle herself, Marissa gradually experiences a change in her mindset once she makes the difficult decision to have surgery. Her attitude in her own words becomes: *Wake me when it's over.* This mindset is common among individuals undergoing surgery. With "experts" now in charge of her care consciously and unconsciously promoting this attitude, Marissa no longer feels responsible for her own treatments and therapies. She mentally hands that responsibility over to her surgeon. She is sending her body into the "shop" for repair. She would rather not know the details or think about it. This attitude is reinforced by our mainstream medical system. No preparation is required of her. "They" are going to take care of her.

Sentient perception and awareness

Sentient perception means being subtly, sensitively, and finely attuned to perceiving minute flickerings of body feelings, movements, sounds, and images that catch your attention.

One handy way to access sentient perceptions is to ask yourself, *What am I not noticing right now or almost noticing or just barely noticing?* Sentient perceptions are at the very edge of our attention. It is also helpful to welcome them by holding yourself open to thoughts and feelings that are irrational and unexplainable, particularly first impressions.

Sentient awareness explores the messages; the dreaming and meaning contained in sentient perceptions; and specifically in Marissa's case, the sentient perceptions in body symptoms with regard to her ankle and impending surgery. These messages are about our lives and are trying to reach greater awareness in us. Until the messages are received and acted on, they will continue to be sent, intensify, and possibly change forms. These messages may intensify via the same body symptom over time, sometimes leading to surgery. They may switch into different body symptoms, or they may manifest as mistakes, accidents, illness, injury, career difficulties, or relationship issues. Working on the messages contained in body symptoms, relieves the body of the burden of carrying the experience alone, by providing awareness and coherent meaning, often resulting in healing.

The healing process

From my experience, I believe there may be ways, using sentient awareness, to enhance the outcome of Marissa's surgery ~ promote faster recovery, reduce pain, and prevent relapse or recurrence. Surgery is an important part of the healing process, <u>but only part of the process</u>, and may complicate the situation in the long term by only relieving the symptoms. Not addressing the whole mental, emotional, spiritual process could invite relapse and recurrence. More complete "healing", as we commonly prefer to think of it, as a long lasting "cure", is a potential outcome for a patient who explores the sentient meaning of their life process being expressed through their body symptom.

In general, there are two types of body symptoms: acute and chronic. Sentient awareness may sometimes activate spontaneous healing of an acute symptom, eliminating the need for other interventions. Sentient awareness may also help relieve and prevent the recurrence of chronic symptoms, although many chronic symptoms are faithful life-long message bearers requiring continuing translation and medical intervention as needed. Marissa's ankle is a chronic symptom that will require deep sentient awareness work.

Marissa's surgery is a month away. With a sense of urgency I tell her it is very important for us to start work as soon as possible. She has no idea what that means. But true to her courageous character, she agrees to meet with me and dive fully into her life.

Marissa's thoughts

Marissa: *I remember being apprehensive and curious about the sentient work. I felt too busy to handle something else. I wanted everything to be fixed ASAP.*

BUT . . . this was a non-western-medicine technique I was unfamiliar with and hadn't tried. What if this is what I overlooked with regard to all other healing alternatives? Should I postpone surgery?

All I knew was you were a master of woo.[1] I just hoped you wouldn't pry into anything . . . what if this brings up inner demons that I am not ready to deal with at this point.

First session ~ the dreambody

In our first session Marissa tells me her history of injuries and treatments. I explain the idea of sentient awareness and the idea of messages and dreams being contained or mirrored in body symptoms and vice versa (Arnold Mindell, *Dreambody: The Body's Role in Revealing the Self,* pp.198-200); I give examples of portals (openings or doorways) we might explore to gain access to awareness; and I describe the possible benefits of the work. I complete an intake worksheet designed to ground and center my own process (Amy Mindell, *Alternative to Therapy,* pp.9-18*)*. I also ask Marissa to do an exercise, to make a quick sketch of herself and a snake.

Marissa is intrigued by our first session and willing to fit in three more sessions, one per week, before her surgery. Each session lasts approximately an hour and a half.

The sketching exercise

My work is based on the pioneering research of Drs. Arnold (Arny) and Amy Mindell, founders of Process Oriented Psychology, also known as Process Work. As dedicated

---

[1] "Woo" or "woo woo" is a colloquial expression for ideas considered non-traditional, alternative to scientific consensus, "irrational".

researchers, their continuing experiential approach is collaborative. They always suggest helpful ideas in their workshops and on their website www.aamindell.net and welcome feedback. I present the following research example/exercise with their permission.

During one workshop, Arny suggests that every hospital admission could be accompanied by a quick sketch done by the patient, of the patient and a snake. His hypothesis is that knowledge of this relationship may help caregivers and patients manage care better.

The snake represents the patient's "unconscious" systems, the autonomic body systems; those involuntary body systems beyond most people's "conscious" control, including the immune system. In addition to the resemblance of the snake to the human spinal cord, vital to many autonomic systems, Arny writes on page 111 of *Dreambody*:

> "What is so vulnerable and special about the serpentine body? The wave-like configuration of the snake represents fluidity, periodic motion, rhythm, and mobility. Wave motion is the opposite of linearity, inflexibility, and rigidity."

I asked Arny if in his experience the snake is the only figure for the exercise, and if so, why? Arny explained, *There is something unique about the articulating serpentine nature of the snake's movement, and the way it strikes so swiftly and accurately.*

Marissa's initial sketch

The patient's own sketch then depicts the relationship between the individual patient and his or her "unconscious" (autonomic and immune) systems, the snake. The key word is "relationship". When I ask Marissa to make a sketch of herself and a snake, this is what she draws.

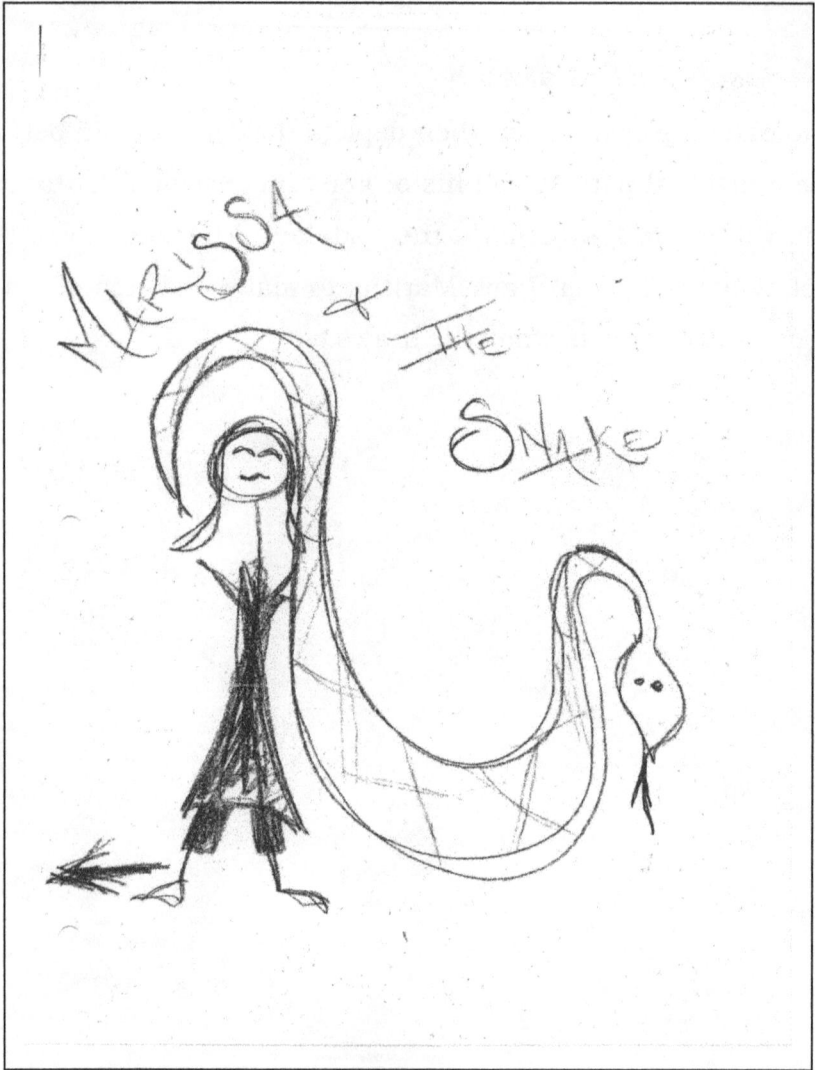

Marissa's Initial Sketch ~ 1ˢᵗ Session

From this sketch, of the "relationship" between Marissa's cognitive process, primary identity, and her "unconscious" autonomic and immune processes, we see disconnect which predicts a slow and painful recovery for Marissa, who is likely to sabotage her own recovery efforts: Her normal cognitive attitude, as depicted by her sketch of her body, is to ignore her "unconscious" autonomic body, the snake, and go off in the opposite direction as indicated by her arrow in the lower left. She will likely continue to push her body beyond its limits during recovery as indicated by her left hand seeming to "straight arm" the snake, and the arrow indicating that she will turn away and attempt to ignore, or attempt to override, her "unconscious" body systems, the snake.

This antagonistic relationship is taxing on the body. The dominant thick snake is obviously in control, not listening to her "ASAP" attitude. Marissa's autonomic body, the snake, is headed in the opposite direction, not looking at Marissa or cooperating with her at this time. There is a separation in progress in this internal relationship struggle which drains energy and taxes the body's healing resources.

Initial prognosis

My prognosis based on Marissa's *Wake me when its over*, "ASAP" attitude, and this sketch, is that without any sentient awareness work she will have less than her optimum healing resources available, and she will likely experience a longer than normal recovery period, higher than average levels of pain, and a higher likelihood of relapse or recurrence.

I describe this sketching exercise in detail not because it is altogether necessary or integral to the sentient awareness work itself, but because it is so graphic in depicting and framing the work. This baseline sketch will be used to demonstrate changes in her relationship to her "unconscious" autonomic body systems as a result of the sentient awareness work we are about to do.

Wisdom of the process

I also remain aware of my bias toward reducing "suffering", which is my hope for Marissa. At the same time I have to remain open to the possibility that "suffering" may indeed be Marissa's path, and that I may have to bow to the wisdom of her process. My bias to relieve "suffering" is further supported because of her openness and willingness to work on herself, which I believe will enhance her recovery.

That being said, "suffering" is in quotes because it represents my outside viewpoint and interpretation of her process, and resonates with and reflects that of our society's medical model. From her viewpoint, inside her process, she may not even experience it as "suffering", but instead as something extremely important, an intense meaningful unfolding experience. For example, as she mentioned, *Physical pain was the least of my worries.*

And this "suffering" may actually be a "short cut", so to speak, in the larger sense of her life. For example in the early hospice movement in the Middle Ages, the shortest path to enlightenment or spiritual healing was considered to be attained through suffering. Suffering was seen to contain a creative, meaningful, and needed message or healing energy. The path away from "suffering" may, in fact, prolong the suffering. This ancient knowledge remains true today, despite our cultural bias toward medications, interventions, and denial which often prolong physical and non-physical forms of suffering and death, especially for many of our elders. So I remain aware of my bias, and also continue work on myself to be open to the wisdom of the process by closely following Marissa's feedback.

In summary, my role as facilitator is somewhat of a misnomer because I do not "make" anything happen. The skill of "following nature" is counter to our Western culture, which is oriented toward "making things happen". Sentient communication facilitation on the other hand follows and supports the flow of nature, the wisdom of the process, the result or side effect of which is often healing.

Feedback

Feedback is any response or communication signal or lack thereof that follows in response to a communication attempt. Feedback can be positive or negative. Following positive feedback changes a process. Following negative feedback maintains status quo. If Marissa had not responded with positive feedback: eagerness, and high energy, to our first session, that would have constituted negative feedback, stasis, and we would not be continuing with the work.

After the first session

Marissa and I never discuss the sketch. She is not curious about it. Her interest is in the experiential work, not in any theoretical discussion. I am now in the role of "surgeon" and she is dutifully submitting to my "doctor's orders", the first of which she has "obeyed" by doing the sketch.

Following our first session, I consult with my friend, colleague, mentor, teacher, and supervisor Stan Tomandl to work on issues that have come up for me in regard to Marissa's work.

Three days later Marissa calls excited to report that she has been dreaming more and has had a "breakthrough" dream. I am pleased to hear about the dreaming because in my experience, it signifies movement, a shift in her process. After only one session, during which she was able to simply open herself to inquiry, self-observation, and the idea of sentient awareness, her "stuck" body processes have begun to express their messages through dreaming. This signifies movement or flow in the dreambody, signaling the beginning of a potentially healing process.

The dreambody

The idea of the dreambody[2] originates from Dr. Arnold Mindell's earliest research work as a Jungian Training Analyst, teacher, and researcher in Zurich, Switzerland. It refers to the relationship he discovered between dreams and body symptoms, the way dreams pattern body symptoms and vice versa; the way they mirror one another.

There is a story about Carl Jung, who near the end of his career was heard to say that his whole approach would

---

[2] Arnold Mindell, *Dreambody: The Body's Role in Revealing the Self.* Boston, Massachusette: Sigo Press, 1982.

have been significantly augmented had he known earlier in his career that dreams could predict illness. Arny's research revealed that dreams can not only predict body symptoms and illness, but can also help prevention and healing, but not in the conventional medical sense.

In short, body symptoms are not only pathological, but are also messengers carrying important information that is trying to manifest as part of a mind/body unity. Healing or relief in the conventional medical sense is often a side effect of following and facilitating this dreambody process to unfold. Arny and Amy Mindell and their colleagues also developed the body of knowledge and skills for doing this known as Process Oriented Psychology, or Process Work.

Confidentiality

Marissa has graciously given me permission to write about her experience. However, I will still limit this discussion to general patterns and skills related to her healing without divulging intimate personal details.

While Marissa's personal sentient experiences are unique to her, the ideas, patterns, portals, and skills employed are universally applicable to other surgical, rehabilitation, medical, healthcare, relationship, family, group, and life situations.

Following and nonlinearity

With only three more sessions before her surgery, Marissa accomplishes months, if not years, of therapy because of her eagerness, openness, curiosity, and amazing courage for self-exploration, and the readiness of the "wisdom of the process".

In other words, it is not simply a cognitive decision on her part or my part that defines her readiness, or the work that we will do. There has been a period of preparation for each of us that neither of us is necessarily conscious of, the wisdom of the process.

And what appears linear in my description of the work is really nonlinear at the sentient level. What I am trying to say is that sentient awareness facilitation is not a "program". It is science and art, linear and nonlinear, cognitive and experiential. And, at all times I am in the role of awareness facilitator following the process as directed by Marissa's feedback and my own sensitivity.

Second session ~ dreamwork

Marissa comes to our second session primed and ready to work on her breakthrough dream. Her dream is the portal (doorway or opening or entry way) that has presented itself as a result of our first session. This week we will gaze into this portal seeking its essence, the sentient experience behind the dream. This dream is about an addictive tendency, which we all have to one extent or another. Our work is to get to the sentient essence behind Marissa's addictive tendency. The tendency is the messenger, not the message.

The dreamwork is emotionally intense and results in a sentient experience for Marissa. By its very nature and the limitations of language, Marissa cannot fully explain what or how she experienced that sentient space, but she is able to name the experience during our session to give it an anchor. The essence or message behind her addictive tendency (deliberately unnamed here to respect confidentiality) involves "Limitless Possibilities" and "No Wrong Turns" (meaning that there <u>are</u> no wrong turns!). Marissa was able to access this altered state of consciousness that her addictive tendency was attempting to access, via a new pattern, an alternate path, without acting out the addictive behavior. She leaves the session highly energized and looking forward to our next meeting. She is also relieved to be past one of her biggest fears, *What if this brings up inner demons that I am not ready to deal with at this point.* She has discovered that demons are messengers, not the message, and she has actually experienced an alternative pattern to her addictive tendency.

Third session ~ moment of injury

In our first session Marissa described a trampoline injury to her right ankle that occurred over three years ago and set this chronic injury in motion. From my experience, I know that returning to the moment of injury can be very healing for the body.

The moment of injury contains a pattern that includes the person's primary or normal identity, and the secondary messages that are seeking awareness. Just as messengers and messages appear in our night time dreams, the same dynamics occur throughout our waking life.[3] In my experience, <u>the body</u>

---

[3] Arnold Mindell, *Dreaming While Awake.* Charlottesville, Virginia:

seems to heal better the more it realizes what has happened to it, and how the incident connects to its process.

The moment of injury is another portal that I am eager to explore. Yes, I confess to an agenda or hypothesis, which is not "following" until, and only if, Marissa responds with positive feedback. She does express interest and energy in response to the idea so we pursue it.

I ask her to reenact the moment of the injury in great detail and in every sensory grounded communication channel (auditory, visual, movement, and body sensation). I have assessed that this direct approach is appropriate for her, in the sense that reliving the accident will not traumatize her because she is able to maintain her ability to observe herself as she works, which she did during the first session.

Again the benefit for Marissa will come from accessing the sentient essence of the obstruction causing the injury at the moment of impact. In this case it is the trampoline. Marissa accesses this sentient experience in an altered state of consciousness and then is able to give it a name to anchor the experience. She names it "Give and Take". This is very relevant to important relationships in her life in the moment. Her sentient experience of this give and take state presents Marissa with valuable insights.

In the fourth session we also work on the relationship of the entire trampoline experience to her childhood myth.

Hampton Roads Publishing Company, Inc., 2000.

Fourth session ~ childhood myth

In our fourth session we work on Marissa's childhood myth. A childhood myth is the pattern that arises from our earliest childhood memory, or our earliest (often recurring) childhood dream. This memory or dream is usually emotionally charged and easily identified. Marissa easily identifies her important childhood dream, a recurring one, and we work on it.

The idea is that the pattern or story in this dream shows up continually during our lives influencing and guiding us along our path. For example, one individual who repeatedly dreamed about being late and unprepared for school as a child goes on to become a school teacher. The school teacher is the "other" or "secondary" figure or energy furthest from her primary identity in her recurring childhood dream. The dream posits "teaching" as her life myth. It does not say how or what or why or give details. It establishes an overall pattern that appears in her life repeatedly. In this instance she grew from primary grade school teacher into curriculum director for her church.

Marissa's recurring childhood dream is more mysterious than the previous example. In her dream an unknown creative source "off camera", out of the frame of the dream, materializes objects which drop through the air then mysteriously land, or not, off camera, outside the other side of the frame of her dream. We can recognize that same pattern in her trampoline accident. On the trampoline, she becomes her dream, the object(s) falling through the air. The trampoline was too tempting an opportunity for her to pass up, the opportunity to unwittingly and physically experience and re-access her childhood dream for a deeper understanding of her life/myth/path. She pushed the outside of the envelope into the mysteries in her myth by jumping higher and landing harder expanding both sides of the "frame" in her myth. Continuing work on our childhood myths throughout our lives peels back layer after layer leading us to richer experience, deeper understanding, and greater appreciation of ourselves.

Another of Arny's discoveries is that our childhood (or life) myth is also related to our chronic body symptoms. Chronic body symptoms can be predicted from our life myth, and working on our life myth can relieve our chronic body symptoms. For example, in Marissa's childhood myth the falling objects are likely to land somewhere sometime and not always gently. So when Marissa takes up the role of the falling object in her dream, it is likely that she may have an "unconscious" landing somewhere sometime, which she did. And her dance career fits the pattern of her myth. Refer back to her picture on page 2.

Fourth session ~ healing essence

At the time I am working with Marissa, Arny and Amy are researching "healing essence", which involves experiential exercises to discover the altered state of consciousness, or sentient essence, that is most healing for our body. So I also offer Marissa the opportunity to do one of these exercises, which she jumps at. She discovers and experiences the sentient state, or essence state, that is most healing for her own body, and names it "exciting anticipation".

The three levels of healing

Healing, as in "complete healing" or "cure" in the conventional medical sense, needs to address three levels of reality: consensus reality, dreamland, and essence.

Consensus reality refers to the "normal" physical world as we typically know it and experience it and agree about it in our consciousness as dualistic, linear, local, and separate. Life is experienced as having a dual nature consisting of perceiver and perceived; as a linear sequence of experiences on a timeline; as a physical location (I am here in this physical location and nowhere else); and as separate (my body and mind are separate from anyone or anything else). This is the primary orientation of conventional medicine.

Dreamland refers to that aspect of the world that is non-linear, non-local, and not separate. We experience this level primarily in dreams, symptoms, and relationships. From physics the term is entanglement, which is also a good description of how we experience this in relationships, symptoms, and dreams. In dreams we experience many different parts of ourselves. These parts or dreamfigures, appear simultaneously, linearly, or randomly, and; dreams move forward or backward in time or out of time, and; we experience ourselves in dreams as both perceiver and perceived, sometimes occupying our "normal" or "primary" identity, and sometimes not separate from, but as someone or something other than our primary identity. In relationships we can "see" ourselves through someone else's eyes, and we can "project" our own growth issues on other people, our family, and groups, thus the term entanglement.

Essence refers to our human need for connection to something larger than ourselves. It is the spiritual dimension in which the world is not experienced as dualistic or marginalized or diversified or entangled, but as unified wholeness through sentient awareness.

These three levels of reality are not to be confused with "body, mind, and spirit" which are consensus reality constructs.

Each time Marissa and I meet I make sure to review all three levels with her (Arnold Mindell, *The Quantum Mind and Healing: How to Listen and Respond to Your Body's Symptoms*, Chapter 2, pp.15-27, "Rainbow Medicine, A Unifying Medical Paradigm"). We discuss her consensus reality level physical treatments and therapies including a few practical suggestions from my own experience. We explore the dreamland level working on relationship entanglements, symptoms, and dreams. And, we work on the deep sentient essence states that address the spiritual healing aspects.

About portals

Body symptoms, dreams, moment of injury, and childhood myth are all portals (doorways, openings, entry points), invitations to explore a deeper creative process that is trying to happen in Marissa's life. All portals lead to the same underlying sentient process in the moment. Childhood myth, however, also predicts chronic symptoms that continue to prod us with messages and signals guiding us in our life path.

Earlier I mentioned that Marissa accomplishes months, if not years, of therapy in a few weeks. The "wisdom of her process" was in high gear, opening and exploring many portals with rapid integration of her work. The width and depth of her work offers this unique opportunity to document and discuss the healing power of sentient awareness for surgery.

Marissa's follow up sketch

At the end of our fourth and last session, the evening before her surgery, I again ask Marissa to sketch herself and a snake. This is what she draws.

Marissa's Follow Up Sketch ~ 4th Session

The resulting change in the relationship between Marissa and the snake since the beginning of our work together three weeks ago is amazing. When we compare her "before" sketch with her "after" sketch, we see a significant change in the relationship between Marissa and her "unconscious" autonomic body systems, the snake. I am greatly relived to see a vibrant, engaged relationship. She is no longer in the process of separation. She is engaged with, aware of, and in balance with her autonomic systems. Marissa and the snake, both of equal stature, are balanced and seeing eye-to-eye. I am confident that her recovery will go well.

Marissa's sketches are unique to her experience and although not really necessary to the work, as I have mentioned, are very graphic in depicting and framing the work. They depict only two of an infinite number of possible relationship patterns. Also, suffice it to say that Marissa is artistic. A more typical sketch for this exercise would involve simple stick figures. No artistic skills are required.

The changes depicted between Marissa's first sketch (p.10) and her second sketch (p.24) resulted from Marissa's sentient awareness work three weeks before her surgery.

Marissa's memories of the work

I have one more piece of work for us to do as soon as possible after her surgery ~ sentient awareness for post-surgical pain management and for processing the trauma of surgery to the body.[4] I will discuss this piece in detail in a moment. But before I do I want to discuss Marissa's memories of the work. I ask Marissa in retrospect, seven years after the fact, what she can remember about the work we did together.

Marissa: *What I can remember about the actual work is that I felt like I was preparing my body emotionally for the surgery. But it is interesting that I do not have any memory of the details of the specific work we did. I do remember drawing a coiled snake. And the most helpful part I remember was working on the pain. That was amazing! I felt like I was able to comfort my body and let it know that everything was okay.*

---

[4] Sentient awareness can also be facilitated during surgery. That would be a cutting edge application of the work, no pun intended.

Altered states of consciousness

Marissa's lack of detailed memories about the work in my office is not surprising despite the fact that she had almost total recall about her decision to have surgery and her anticipation of working with me. This is because during each of our work sessions Marissa, in accessing her deep sentient nature, was in and out of altered states of consciousness.

Altered states of consciousness are states of being that we do not recognize as "normal" ranging from something as mild as day dreaming or feeling disoriented (not myself) at one end of the spectrum, to conditions such as delirium, dementia, and deep coma. We all experience altered states of consciousness throughout the day and night. Although our society treats them as abnormal for the most part, they commonly occur, or "normally" occur, every day and night in every culture throughout the world. Sentient states of consciousness are experienced as altered states of consciousness.

Altered states of consciousness are like new dimensions, parallel universes, and are often the quickest route to the deep new creative sentient messages unfolding in our lives. But like parallel universes, without specific training, most people experience them as separate places, separate experiences where memories from one universe do not automatically translate or transport to the other universe. In Marissa's "normal" state of consciousness she cannot always access memories from her "sentient" altered states of consciousness, and does not need to.

Also, working on deep sentient awareness takes us into altered states of consciousness where the cognitive process plays a lesser role, and may not even be required for integrating the messages/experiences.

The surgery

I meet Marissa at the hospital at 8:30am Tuesday morning. Marissa's outpatient surgery is scheduled for 10:00am. The surgery actually commences at 12:30pm and lasts until 2:30pm.

The surgical procedures include; right ankle arthroscopy (endoscopic joint surgery), removal of a loose bone fragment, modified Brostrom lateral ligament reconstruction, and repair of the peroneal tendon rupture. Suffice it to say, this is extensive ankle surgery.

The anesthesiologist recommends general anesthetic although other local alternatives are available. The anesthesiologist assumes without asking that "wake me when it's over" is the preferable choice of young adult patients.

In retrospect general anesthetic, although in some instances more wearing on the body's resources and awareness, was the best alternative in this case, because the procedure turns out to be more extensive than anticipated.

Family waiting room

At 3:00pm the surgeon emerges and explains that the surgery was more extensive than he thought it was going to be, particularly the bone grinding. We see pictures of the complete lack of stability in the ankle prior to the surgery, pictures of the damaged and inflamed tissue that was removed, and a picture of the completely stable reconstructed ankle.

Although this is supposed to be an outpatient procedure, he strongly recommends hospitalization for pain management. He says she can expect five days of pain. His tone implies pain with a capital "P", which he explains is because of the extensive bone surface grinding.

Recovery

There is no visitor access to the recovery room from 2:30pm to 4:30pm. Marissa is released to her room at 4:30pm . We see her approximately two hours after she gets out of surgery.

Marissa is in excruciating pain and being administered a narcotic drip. Staff agrees to increase the dose. They give her a button to self-administer her own meds as often as every eight minutes.

Sentient awareness for post-surgery pain

In my experience, I know that working on sentient awareness as soon as possible after trauma can accelerate and enhance the healing process. I jump in at the first opportunity.

Tom: *What color is your pain?*

Marissa: *Purple.*

Tom: *What shape is it?*

Marissa: *Long narrow cylinder.* She motions up and down with her right hand. *It's very dark.*

Marissa is under the lingering effects of general anesthesia, a narcotic drip, and fading off to sleep. And, I am too exhausted after nine hours of waiting and anticipation to do more. But even with this minimal piece of work Marissa has already experienced her pain in the visual channel ("purple" and "long narrow cylinder" and "dark") and the movement channel (up and down with her right hand) which gives both of us more information about the pain, and based on my experience provides momentary relief to her body from the burden of carrying the experience alone.

Supervision for the sentient care™ facilitator

I return at 9:00am the next morning. Marissa is off the narcotic drip and on oral narcotics. She is in severe pain and the doctor agrees to test different medications to manage the pain, alternating two oral types, one narcotic and one non-narcotic.

Marissa: After a long pause she comments, *It feels like three bees stinging right on the line of the surgery.*

Tom: I venture, *Why are they stinging?* Even as the question leaves my mouth I realize I am attempting a poor intervention by leading with a question instead of a statement to support her experience in the moment.

Marissa: *I don't' know.* This is negative feedback; a dead end.

Minutes later as the pain remains intense and the doctor is considering alternative medications, I realize Marissa is still in a good state of consciousness to work on her awareness. I try to pick up the thread from her last signal.

Tom: *Be a bee.* This is a no go. No positive feedback comes from Marissa. With that I realize I am unfocused. I go out for coffee and call Stan for supervision. I come back grounded in my awareness skills, ready to work with her.

Following Marissa's process

I return to work with her at 11:30 am.

Marissa: *Now it feels like someone with one of those little plastic mallets the doctor uses to test your reflexes, pounding right on the bruised spot of the surgery.*

I start tapping the air with one hand. Marissa corrects me.

Marissa: *No, it's two mallets.*

I start tapping the air with two mallets. I continue tapping away. Marissa corrects me again.

Marissa: *No, it's got a rhythm.* Following her feedback, I break out into a drummer's routine and calypso dance. Marissa gives positive feedback, a big smile and a laugh.

Marissa: The process evolves. *Now it feels like a flow of pain like an egg cracking and flowing.* I ask her to demonstrate with her hands. She uses her right hand to slowly crack an egg over her left hand. I ask her to repeat it. She does the movement then changes the subject.

Marissa: *Now the bees are back, but there are only two of them.*

Tom: *You are a bee. Who is with you?*

Marissa: *A friend.*

I role play Marissa. Marissa role plays the bee.

Tom: *You are stinging me.*

Marissa: *I am just imaging how gross it would be to be a bee and to be stinging a soft puffy purple bruised swollen place . . . . Well, maybe it would be a soft place.*

Tom: *You fly in here and do that and then leave, and come back later and do it again. What happened to the other bee?* I could have followed the "soft place", but I am guessing that the other bee is even further from her awareness making it possibly more sentient.

Marissa: *I guess he had better things to do . . . Now the breaking egg is back.*

The chef of pain

I ask her to repeat the egg breaking hand movements several times.

Tom: *What is the egg breaking on?*

Marissa: *Bone.*

Tom: *How is the egg breaking? Is it being thrown, dropped, held?*

Marissa: *It is being broken very skillfully.*

Tom: *Sounds like a chef . . . maybe a master chef.*

Marissa: *It is THE CHEF OF PAIN!* She exclaims with a great big smile from ear to ear.

Tom: I repeat, *YES, it is THE CHEF OF PAIN!*

Marissa: *Oh . . . I don't have any pain in my leg. I don't know if the medicine is kicking in or what . . . Look, I can wiggle my toes without any pain!*

Tom: *It is because you have done such good work following your sentient awareness. When you become the Chef of Pain it relieves the body.*

Marissa cannot hold this state for more than a minute.

Marissa: *I don't want to work any more right now.*

Tom: *Yes, it is time for a break.* I leave and take a two hour break. When I return to the hospital at 2:00pm her mother says: *Her pain was much better after you left.*

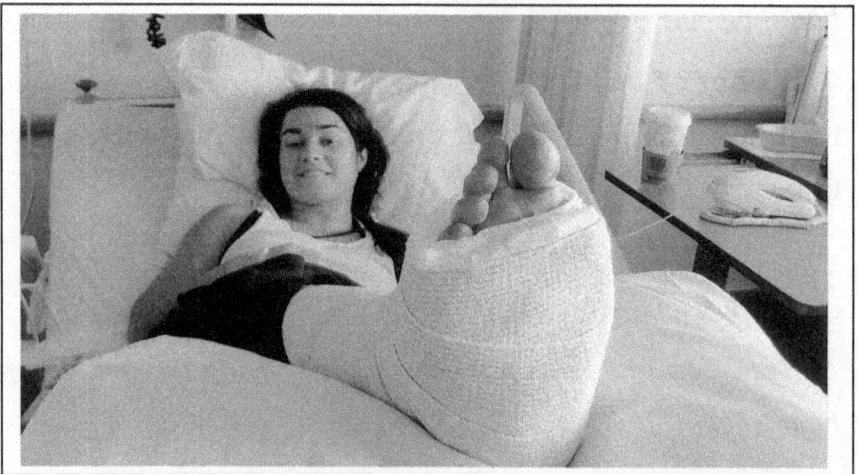

Marissa in her hospital room after sentient awareness work

Home Again

The following day after her surgery we pack her up and arrive at her home around 3:30pm. Lowering her elevated foot to get into the house on crutches and switching to a new oral narcotic prescription has amplified the pain. We settle her in with ice packs, pillows, remote control, food, and chocolate.

At 5:00pm, before I leave for dinner, I try one more intervention picking up where we left off at the hospital. I try an intuitive leap.

Tom: *You told me that The Chef of Pain is very skillful. Maybe The Chef of Pain is trying some new dishes.*

Marissa: *Yes, and it is definitely not an egg dish this time.* Positive feedback.

By 11:00pm Marissa has transitioned into recovery mode and has taken charge of her own life by taking control of her immediate environment and dismissing any more help.

The following morning, Wednesday at 10:00am her mother reports with some surprise and astonishment in her voice: *Marissa had a pain free night.*

Summary

Because of the substantial number of nerve endings in the foot and ankle, surgery there is one of the most painful procedures to undergo. Marissa's surgeon predicted that she would experience five days of above average levels of pain, and strongly recommended hospitalization post-surgery for pain management.

However, despite an extensive surgery including substantial bone surface grinding, which is excruciatingly painful, Marissa experienced a pain free state momentarily the day after surgery, and a pain free night within 48 hours of surgery. When new information and meaning from pain has been accessed, a brief pain-free moment is enough in most trauma cases including accident, illness, and surgery, to accelerate the healing process.

Marissa was pain free the second night after surgery and experienced lower than anticipated levels of pain subsequently, thus enabling her to reduce and then discontinue the oral narcotic painkillers in far less than the anticipated time. In my experience, if less pain killers are needed, the patient's awareness of their own healing patterns and energy will be more active and the body will heal more rapidly.

Marissa fully recovered months earlier than anticipated. Her surgeon expressed his surprise. He is the official surgeon for another major Chicago dance company and thoroughly familiar with Marissa's type of injury. He stated that he has never seen a surgery as extensive as Marissa's heal as rapidly and successfully as Marissa's. He attributes this to his own skills, skills that are truly admirable. However, the one salient difference between Marissa's recovery compared to that of his other patients is the sentient awareness work we did before and after Marissa's surgery.

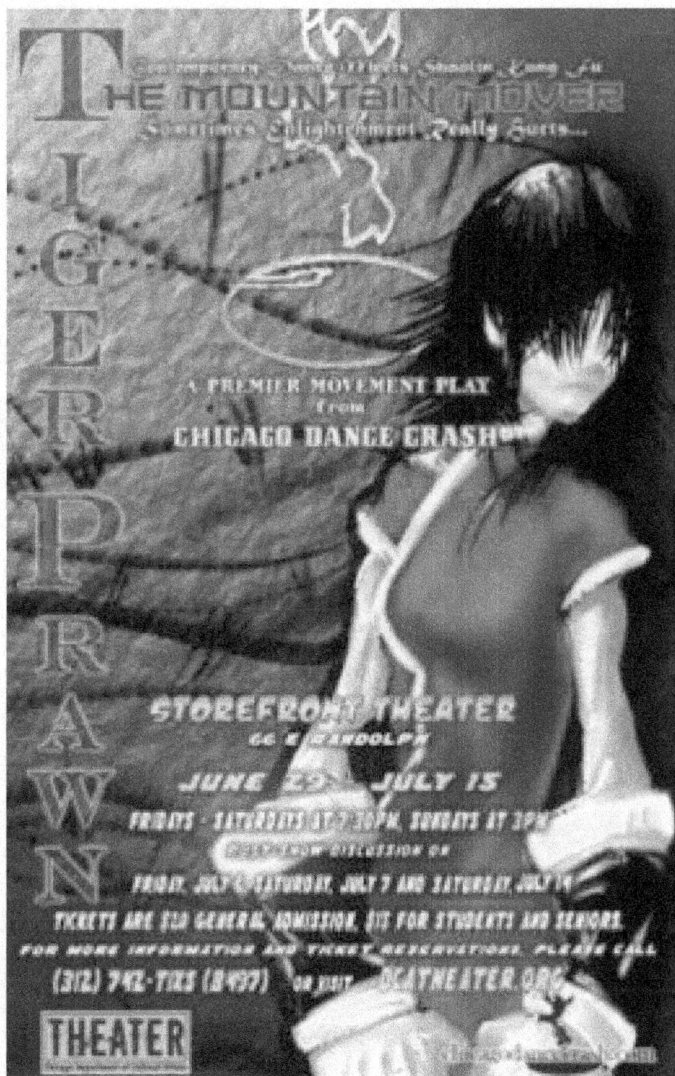

Poster art and design by Kyle Vincent Terry

Tiger Prawn: The Mountain Mover

Epilogue

Marissa enjoying Tiger Prawn "afterglow"

Post surgery, Marissa's professional dance career includes the title role in *Tiger Prawn,* the Chicago Dance Crash company's action-packed martial arts extravaganza; and a return to her Latin roots, touring nationally with the Chicago Latin Street Dance Company.

Photograph by Lindsay Schlesser

Marissa ~ Tiger Prawn ~ Front Claw

Photograph by Andres Meneses

Marissa ~ Chicago Latin Street Dance Company

Bodywork exercise

Thanks to Drs. Amy and Arnold Mindell for developing this bodywork technique.

Just as an ounce of prevention is worth a pound of cure, sentient awareness bodywork on a regular basis can sometimes prevent the need for surgery. And in the event of surgery, sentient bodywork can promote faster recovery, diminish pain, and help prevent relapse or recurrence. Here is a sentient bodywork exercise:

1. Think of the body problem that bugs you the most. Notice the energy of that body symptom. For instance tight, throbbing, sharp, dull, burning, cold, etc.

2. Gaze around the room until you see something that catches your attention and reminds you of that energy.

3. Look at it and discover why it flirts with you; why it catches your attention. Notice its qualities.

4. Notice who inside you really needs to see that energy right now. (Note: That energy has probably been trying to get your attention all day.)

5. Who set up the whole scene: the body symptom; the flirt; and the observer? Who is the writer, choreographer, director, and producer of the whole scene? (This could be a spiritual figure, a figure from

mythology or fairytales, a movie character, or someone from your past, etc.)

6. How can you use this energy to navigate your day: relationships, work, leisure, etc.?

Margaret and Rehabilitation

Margaret, age eighty-five, is in a rehabilitation center recovering from an excruciatingly painful hairline fracture of her sacrum, probably related to a fall. After physical therapy to strengthen her walking and improve her balance, she is resting in bed. She describes some "problems" she experienced during her physical therapy today. I suggest doing this body work exercise.

Tom: *Think of the body problem that bugs you the most. Notice the energy of that body symptom.*

Margaret: *I feel soft pads on the balls of my feet . . . like cotton balls . . . more on the right than the left, but on both feet. I am afraid my feet will go numb, and I won't be able to walk.* In addition, Margaret is afraid that she will not be able to return home. She has to pass a physical therapy test to earn the privilege to return home. And failing that, she fears being nursing home bound for the rest of her life.

Tom: *Gaze around the room until you see something that catches your attention and reminds you of that energy.*

Margaret: *I got it right away . . . the shoes on the dresser . . . my shoes never have fit me . . . my shoes have never been comfortable.*

Tom: *You have never been comfortable in your own shoes.*

Margaret: *I get it . . .* **Yeah, I get it!**

Tom: *Notice who inside you really needs to see that energy right now.*

Margaret: *The Tense One.*

From this significant moment of insight, I can see that Margaret has had a shift in her body energy: eyes down, relaxed shoulders, stillness, and slower lower abdominal breathing, as a result of her new awareness. I could expand the exercise at this point to facilitate a more cognitive integration of this new awareness by asking her, *How can you use this energy to navigate your day: relationships, work, leisure, etc.?*, but integration (acting on this new awareness) does not always have to be cognitive. And a cognitive intervention can also be an interruption of a sentient state, which is what I sense it would be in the moment. So I follow the state she is in as described above which is a non-cognitive moment of sentient awareness integration, or grounding, that is enough to unleash her inherent energies and make them available for her life, living, and healing.

Tom: *Who set up the whole scene: the body symptom; the flirt; and the observer? Who is the writer, choreographer, director, and producer of the whole scene?*

Margaret raises and opens her arms tilting her face upward indicating the space above her, toward a power greater than herself.

We visit a short while longer before I take my leave.

The following morning I call her.

Margaret: *Do you want to hear a miracle?*

Tom: *Sure . . . of course.*

Margaret: *Last night I woke up in the middle of the night and felt my feet. They were fine. They were healed! . . . and I had the best sleep I can remember. I even rolled on my side and slept like a baby . . . and I dreamed for the first time in a very long time.* - Rolling on her side had not previously been possible with this painful fracture. Note that not only was the presenting body symptom healed (the "cotton pads" on her feet), but also her excruciating back pain was relieved.

This is a dramatic example of the healing power of sentient awareness for an acute body symptom. An example of how simple, gentle, and effective it can be. However, it must be said that Margaret was open, receptive, and ready to integrate this awareness. The wisdom of the process is always present. I was able to facilitate this outcome with this exercise in only a few minutes, because the outcome itself was already trying to happen through the presence of the body symptoms in her feet. By using this exercise, I have simply helped unfold Margaret's sentient awareness of her inherent life energies.

Who is walking with Margaret?

A week later Margaret is lying in bed, still in the rehabilitation center, after more physical therapy to strengthen her walking and improve her balance. She passed her test last week, but elected to stay for one more week of physical therapy.

Tom: *Can you describe one of the "problems" with your walking?*

Margaret holds her two hands in front of her while she describes and demonstrates how her toes point slightly inward as she walks. She moves each hand in turn demonstrating the walk with her hand movements. I encourage her to repeat the movements several times.

Tom: *Who or what is it that moves like that, or makes that movement?*

Margaret: *A duck.* Her response is instantaneous with a strong sense of recognition. No doubt about it. It is a duck.

Tom: *What kind of duck?*

Margaret: *A waddling duck.*

Tom: *Where is it waddling?*

Margaret: *To the ocean . . . Oh! . . . It's Donald Duck!*

Her recognition and knowledge of the dreamfigure is so strong; I go right for the meaning.

Tom: *What is the essence of Donald Duck?* I don't even have to explain "essence" or give examples. She knows it instantly.

Margaret: *Humor! . . . and ducks have good balance.*

Tom: *What is humorous?*

Margaret raises her right hand, thumb by her mouth, fingers extended. She curls her fingers, sticks out her tongue, gives me the "raspberries" (blows out), and bursts out laughing. She has taken me completely by surprise. When I recover momentarily after being startled, I burst out laughing too.

Margaret: *See . . . You laughed.*

Tom: *You need humor!*

Margaret: *That's right! . . . I've been too tense.*

Analysis

Initially I use the every day (consensus reality) word "problem" to inquire about those aspects of Margaret's walking that are not walking with her; those aspects that are causing her difficulty walking. These "problems" are actually manifestations of the dreambody containing messages that are important to Margaret's overall life and well being. These "problems" or body "symptoms" are dreams trying to happen in her physical body. "Someone", a dreamfigure with a message, is always walking with each of us, and if we are curious enough we can seek it out and ask its advice (or essence).

In this example we begin with a sentient movement. Notice that Margaret does not actually have to get up and walk to explore walking. The sentient movement is the slight turning in of the toes that she was able to show with her hands.

Exploring this sentient perception uncovers the dreamfigure Donald Duck. And the message Donald Duck has for Margaret is, *Humor! . . . Lighten up!*

"Donald Duck" (her slightly turned in toes) is likely to stay faithfully with Margaret until she thoroughly gets the message and lightens up despite the best intentions of physical therapy to "straighten" the position of her feet. Physical therapy (PT), while vital and necessary to Margaret's well being is simply not the whole story. PT is not treating Margaret's "dis-ease". Both PT and sentient awareness are necessary if she is to regain her balance and maintain it.

She identifies with me laughing at her antics, but she has not yet completely identified with her own laughter and humor. That is what she needs to work on and play with. Knowing the message is a great start and can bring temporary relief, but the message must be integrated, acted on, to help relieve the body of the "problem". More than likely her toes will straighten out if she actually can lighten up.

Symbolically she is "out of balance" because she is "too tense" all the time. Humor is what her own dreambody is prescribing as the next step to putting her walk and life back in balance.

Humor is the natural remedy, as opposed to the medical allopathic pharmacologic approach; although drugs are certainly widely used and highly touted for "lightening up".

Charlie Brown goes to the beach

For the last year, after returning home from rehabilitation, Margaret has been living independently, and maintaining her own home. During this time she has had a recurring symptom, her "stomach ache". A week earlier she went to her doctor who prescribed a pain medication. She obediently took it, and has been nauseated and throwing up for the entire past week. Today she returned to the doctor for the prescription that "cured" her last year at this time. She expected an argument from her doctor, but they were able to identify her old prescription, and she is hopeful that it will again affect a "cure".

I call Margaret to get an update on her condition, and the outcome of her doctor visit. I suggest that we try working on her body symptom in the moment over the phone.

Margaret: *My stomach hurts.*

Tom: *Where exactly does it hurt?*

Margaret: *In the center of my body . . . at my waistline.*

Tom: *Behind your belly button?*

Margaret: *Between my belly button and my ribs.*

Tom: *Okay. Notice your stomach ache closely, without making yourself too uncomfortable . . . (Pause) . . . What color is it?*

Margaret: *Flesh colored.*

Tom: *Now notice what shape it is.*

Margaret: *It's the shape of a pomegranate. I don't know why I said pomegranate.*

Tom: *Is there any sound that goes with it?*

Margaret: *No sound.*

Tom: *Okay. Now notice the energy of this flesh colored pomegranate, and make a motion with either empty hand to go with that energy.*

Margaret: *Okay, I am making circles with my left hand*

Tom: *Good. Now notice who is against the pomegranate energy.*

Margaret: *My mind . . . I want it to go away.*

Tom: *Okay. Now put your phone in your other hand . . . Notice the energy of your mind, and make a motion with your other hand, your right hand, to go with that mind-energy.*

Margaret: *It's a kaleidoscope. My hand is moving in all directions.*

Tom: *Good. Now go to your favorite earth spot . . . Your favorite place on the earth.*

Margaret: *I'm there . . . the beach.*

Tom: *Now become that earth spot . . . become the beach . . . the sand . . . the water . . . the wind . . . all of it . . . (Pause).*

Margaret: *Okay . . . I'm the beach.*

Tom: *Now, while you are the beach, put down the phone. With each hand make the two energy movements, the circles with your left hand, and the kaleidoscope with your right hand. As you are the beach, watch them as they move and interact . . . (Pause) . . . What happened?*

Margaret: *They became the ocean waves.*

Tom: *Take a moment to notice your stomach ache.*

Margaret*: It's a little better.*

Tom: *Notice the flesh colored pomegranate, and tell me if it resembles anybody.*

Margaret: *It's got two big round eyes . . . and puffy cheeks. . . Oh, it's Charlie Brown . . . !*

Tom: *Great. Now invite Charlie Brown in.*

Margaret: *You want me to send Charlie Brown to the beach?* It is fascinating that her "mind-energy" still wants to banish the body symptom, instead of integrate it or learn from it, so much so that she doesn't hear what I said.

Tom: I express my impatience. *That's not exactly what I said.* Then I realize that my last communication was not very clear, and that I am too exhausted to "follow" her process. "Proper form", so to speak, for doing sentient awareness work would be to support Margaret in sending Charlie Brown to the beach.

Margaret: *Oh . . . I thought you wanted me to get rid of Charlie Brown.*

Tom: *No . . . That's your "mind-energy". You need to take Charlie Brown with you to the beach.*

Margaret: *I'll call my Uncle Dan and see if he wants to go to the beach too. Is an apartment available . . . ?*

Tom: *Yes, probably . . . Notice your stomach ache. How does it feel?*

Margaret: *Oh, it feels better.* Just Margaret's thought of taking action toward the "beach" creates a healing energy flow.

Discussion of continuing work

The work ends where it does because I run out of energy from "pushing" for too long against the wisdom of her process. I have a goal of healing her quickly. When in actuality her chronic symptom is a life process: the mythic battle between her "stomach ache-energy" and her "mind-energy".

Continued work could include having her role play Charlie Brown, and having her spend more time in the altered state of the beach and the waves, with the energies interacting and flowing or transforming. These represent two different levels. Working with Charlie Brown is working with the dreambody at the dreamland level. Working with the energies from the state of the earth spot is working at the essence level.

One year ago Margaret's dreamfigure was Donald Duck, who was also walking to the ocean. And Donald Duck's message for Margaret was, *Humor; Lighten up!* The follow up to today's work would be to unfold Charlie Brown's message further. We can speculate, of course, that it might include the characteristic words, *"Good grief, Charlie Brown!"*

In lieu of continuing the work, I give Margaret a prescription. My prescription is to actually go to the beach, and to work on the Charlie Brown message further by "taking Charlie Brown with her to the beach." "Taking Charlie Brown to the beach" is a "natural" remedy and potentially a permanent "cure", as compared to the conventional medical allopathic pharmacologic approach. Margaret is still anticipating her "cure" from her psychotropic drug of choice, which is something to make her in her own words *"unwind when everything is up tight."* Notice that the drug may relieve or "mask" her "physical" symptom, like aspirin-type products for a headache, but it does not address her underlying disease, Margaret's need to find a pattern for "lightening up". In this sense, the drug is a short-term solution, and perhaps all she needs in the moment, but is not the more permanent "cure".

The body symptom relief that Margaret experiences during the work is likely temporary until she integrates Charlie Brown's message into her life, and experiences the two energies flowing together, which she did experience and gain a pattern for momentarily during the exercise (the motion of the waves in the movement of her hands). The relief of her stomachache in the moment comes from experiencing her body symptom in other communication channels; including visual channel (color and shape), inner relationship channel (Charlie Brown), and movement channel (the flow of the two energies using the movement of her hands). By experiencing the body symptom in these other channels, the symptom becomes "unstuck", and the body channel is relieved of the burden of carrying the message alone, which temporarily relives the symptom.

Margaret's last two exercises above, a year apart, show both a gradual shift in the nature of her process, from "Donald Duck" as the dreamfigure to "Charlie Brown"; and from the excruciating pain of a fractured sacrum to a mild stomach ache. And they demonstrate how long a message may take to unfold (as in years). Margaret has still not fully embodied "the beach" and the "waves" ~ as in get to the "beach state of mind", as in get on "island time", as in let go of "The Tense One" that still inhabits her. And she needs to flow more like the waves. Also note the similarity of Margaret's wavelike movement with her hands to that of the snake's wavelike movement discussed earlier in Marissa's work. But Margaret is now flowing more between the states compared to a year ago when "The Tense One" had her flat on her back in rehab. This "flow" between the energy states is what diminishes the likelihood of relapse or recurrence.

Caring for the caregiver/facilitator

The fact that I am exhausted from the work indicates a need for me to work on myself. If I were "flowing" with the work I would actually be energized from it, not exhausted. Areas to work on might include my impatience, my high expectations, my relationship with Margaret, and my over investment in the caregiver/facilitator role ~ to the detriment of my own need to care for my own well being before I get exhausted. All this makes me "The Tense One", and I empathize with Margaret who became "The Tense One" for much the same reason, because she was the sole caregiver of her ailing husband for the last ten years of his life.

This is also an example of entanglement. The Tense One is a role in the field between Margaret and I, which can be occupied by Margaret, or by me, or can remain unoccupied. An unoccupied role, or "ghost role", disturbs the field by seeking awareness. Its message is encoded in signals including tension and body symptoms, such as my exhaustion.

Facilitated versus un-facilitated sentient awareness

This book describes facilitated sentient awareness with a chronic body symptom requiring ankle surgery, an acute body symptom healed with a bodywork exercise, and chronic symptoms during rehabilitation.

Having a facilitator is indeed a privilege and usually accelerates the work, but it is also very useful to learn to work on yourself alone. The "Bodywork" exercise on page 39, the "Who is Walking with You" exercise on page 44, and the "Favorite Earth Spot Energies" exercise beginning on page 47 that Margaret did can all be done on your own. These are examples of ways to start working on your own body symptoms. Always work to your own comfort level.

Process Oriented Psychology ~ Process Work

Process Work is one very good method for facilitating sentient awareness as demonstrated in this book, and supports and encourages other methods and schools that lead to deep sentient awareness, whether facilitated or done as inner work on your own. For residential and at-a-distance programs visit the Process Work Institute Graduate School, Portland, Oregon, USA at www.processwork.org.

The healing power of sentient awareness

If sentient awareness could be bottled it would be heralded as a "miracle drug" for the body. The more a person can be aware of their process, the more resources the body can draw on to "heal". The reduction of pain is an important goal, and very often a secondary result of sentient awareness work, which also often promotes healing and prevents relapse or recurrence. Sentient awareness interacts in both the macro and the cellular realms of the body. Sentient awareness does not replace medical care. Sentient awareness is an act of reverence that compliments and enhances the whole person; mind, body, and spirit.

The skeptics

Skeptics may argue that this approach seems to require an inordinate and impractical amount of time and training compared to conventional alternatives.

The counter arguments are that 1) Sometimes it can take less time than traditional treatments and, in fact, can eliminate the need for other therapies, and 2) There are little or no "side effects" compared to the inherent side effects of drug interventions and the "scarring" of surgery, and 3) More time invested up front can save many times over the cost of prolonged recovery or relapse or recurrence, and 4) The basic skills and benefits can be learned with a modest amount of training. As in any discipline, mastery, however, is a lifetime endeavor.

Skeptics who do not see this approach as a panacea are absolutely correct. It is not a "program" that can be simply implemented. It is a paradigm shift that offers significant health benefits and economic savings to a large population. I encourage healthy skepticism, and more research along these lines, which can only help develop the work.

The case for a major healthcare breakthrough

A major breakthrough for the world's health care crisis is not going to come only from increasing health insurance for more people, or from implementing electronic medical records, or from quality inspections, or from financial restructuring, or from new drugs, or from "conventional" medicine, or from "alternative" medicine or from "science" as we know it.

These are all aspects of the predominant allopathic healthcare paradigm, a dualistic paradigm rooted in consensus reality that has come to mean that all body symptoms, illness, and disease are the result of something "wrong" or "pathological", caused by "bad agents" and must be eradicated. A major healthcare breakthrough from within this existing paradigm is unlikely. To quote Einstein, "A problem cannot be solved from within the consciousness that created it."

A major breakthrough will come from a transformation of the healthcare paradigm. One possibility is the shift to a process paradigm ~ with more healers versed in multi-level care, which includes the healing power of sentient awareness, especially for chronic conditions; and more individuals re-awakened to their mortality and their responsibility for their own health, actively engaged in all levels of their process, as Marissa and Margaret and I demonstrate here.

This new paradigm would embrace and transform the concept of good health from "balance" or "wholeness", a "static state" and common goal of many contemporary paradigms, to "process" meaning "the flow of nature", which not only includes balance, but also the dynamic flow; the linear and non-linear, the local and non-local, the separate and entangled, and the spiritual realities of health.

It would transform the concept of healthcare to include matching the "levels" of treatment (see "The three levels of healing", pp.21-22) to the levels of the patient's process, including symptoms and beyond, to attend to the underlying dis-ease, entanglements, and sentient awareness. This new paradigm could unleash the incredible and mysterious healing processes inherent in the human body, and make them more available for life and living and healing.

*   *   *

References

Diamond, Julie and Spark Jones, Lee. *A Path Made by Walking: Process Work in Practice.* Portland, Oregon: Lao Tse Press, 2004.

Goodbread, Joe. *The Dreambody Toolkit: A Practical Introduction to the Philosophy, Goals, and Practice of Process Oriented Psychology.* Portland, Oregon: Lao Tse Press, 1997.

Goodbread, Joe. *Living at the Edge: The Mythical, Spiritual, and Philosophical Roots of Social Marginality,* Portland, Oregon: Nova Science Publishers, Inc., 2009.

Mindell, Amy. *Alternative to Therapy: A Creative Lecture Series on Process Work.* Portland, Oregon: Lao Tse Press, 2006.

Mindell, Amy. *Meta Skills: The Spiritual Art of Therapy.* Tempe, Arizona: New Falcon Publications, 1995.

Mindell, Arnold. *Dreambody: The Body's Role in Revealing the Self.* Boston, Massachusetts: Sigo Press, 1982.

Mindell, Arnold. *Dreaming While Awake: Techniques for 24-hour Lucid Dreaming.* Charlottesville, Virginia: Hampton Roads Publishing Company, Inc., 2000.

Mindell, Arnold. *The Dreammaker's Apprentice: Using Heightened States of Consciousness to Interpret Dreams.* Charlottesville, Virginia: Hampton Roads Publishing Company, Inc., 2001.

Mindell, Arnold. *Quantum Mind and Healing: How to Listen and Respond to Your Body's Symptoms.* Charlottesville, Virginia: Hampton Roads Publishing Company, Inc., 2004.

Mindell, Arnold. *Working with the Dreaming Body.* London, England: Arkana, 1989.

Morin, Pierre. "The Dreambody: A New Integrative Approach to Illness". Marylhurst, Oregon: *New Connexion*, July/August 2003. ~ available at www.newconnexion.net/article/07-03/morin.html

Tomandl, Stan. *Coma Work and Palliative Care: An Introductory Skills Manual for People Living in Delirium and Coma.* Victoria, Canada: Coma Communication, 1991.

Glossary [1]

allopathic medicine ~ The predominant medical paradigm where body symptoms, illness, and disease are considered to be something "wrong" or "pathological"; caused by "bad agents" that must be eradicated. Originally used to differentiate conventional medicine from homeopathy. Evolved in common use to mean the practice of Western medicine.

altered state of consciousness ~ An unfamiliar mode of perception. Any condition other than our normal identity, including both our normal waking condition and ordinary restful sleep. Altered states range from mildly tranced out as in day dreaming, or "I am not feeling like myself today"; to conditions such as delirium, dementia, and deep coma. Altered states provide opportunities, and can be "short cuts", to discover new information.

autonomic nervous system ~ That part of the nervous system below the level of consciousness where most activities are involuntary, such as heart rate, digestion, breathing.

channel of communication ~ The mode of perception in which information is sent and received. Sensory grounded channels include: visual, auditiory, body sensation (proprieceptive), and movement (kinesthetic).

coherence (coherent meaning) ~ The feeling of identity, familiarity, order, belonging, connection, security, and alignment in our health and life. It is derived from our compelling human need to make sense out of our own corner of the universe and everything in it, including our relationship to it and our place in it. Depending on cultural and personal beliefs, a symptom or diagnosis may disturb or add to our sense of coherence. One person's sense of coherence does not necessarily "make sense" from other points of view.

consensus reality ~ The "normal" physical world as we typically know it and experience it and agree about it; as dualistic (consisting of perceiver and perceived), linear, local,

---

[1] Definitions of Process Work terms as they are used in this book.

and separate. In Process Work one of the three "levels" of reality.

deep democracy ~ The philosophical basis of Process Work in which all voices, states of awareness, communication signals, and levels of reality, both central and marginal, both outer and inner, are considered important and necessary; and are represented and recieved with openness to their potential creative wisdom, which arises from the continuing relationship flow of the process.

dream ~ The signals, disturbances, and subtle tendencies that appear spontaneously and involuntarily during sleep, lucid, and awake states; and are experienced initially as "other", or "not me".

dreamland ~ That aspect of the world we experience both as perceiver and perceived in dreams, symptoms, and relationships that are entangled, non-linear, non-local, and not separate. In Process Work one of the three "levels" of reality.

dreambody ~ Dr. Arnold Mindell's discovery of the way dreams pattern or mirror body symptoms and vice versa. Body symptoms are not only pathological, but are messengers carrying important information that is trying to manifest as part of a mind/body unity.

dreamfigure ~ The people, animals, figures, objects, events, symbols, and symptoms that appear in dreams.

eldership ~ The nature of an individual, irrespective of age or role, or the nature of a role in a given situation that 1) includes a degree of detachment from the drama; 2) excludes attachment to any agenda, hidden or otherwise; 3) and through a transparent connection to the deepest self creates sacred space, a loving and supporting atmosphere through openness, and acceptance of all parts.

essence ~ The original property or root of an experience that came before it became an expereince; before anything was agaisnt it; the unified, undivided, unpolarized impulse, tendency, or characteristic. Example in the book: What is the essence of the experience of Donald Duck as a dreamfigure? Margaret's experience, *Humor* (p.44).

essence level ~ The unified, unbroken wholeness out of which signals, dreams, and all other experiential phenomenon arise. The deep spiritual dimension in which the world is experienced as unity - not dualistic or marginalized or diversified or entangled. In Process Work one of the three "levels" of reality.

facilitation ~ The eldership practice of deep democracy; to create sacred space, and provide for all parts of a process to be noticed and represented and related to.

feedback ~ A signal of response (communication signal or lack thereof) to an input, that further reveals the nature of a process. Feedback can be positive or negative. Following positive feedback changes a process. Following negative feedback maintains status quo.

following ~ A Process Work metaskill. The skill of "following the deepest flow of nature" - which runs counter to our Western culture that is oriented toward "making things happen". Following includes, in part, noticing the flow of the process as it evolves through and between the communication channels and levels. For example, an observer/facilitator tuned to only the auditory and visual communication channels would experience something like a movie trailer, and miss the multi-level flow. They will see the trailer and miss the movie. Following includes not just following the patient, but also following the facilitator and all parts of the field of the process. For example, the process of the role of The Tense One flows in the field between patient and facilitator in Margaret's example (p. 53).

ghost role ~ A polarized role and unidentified dreamfigure in the field of a process, not yet occupied by anyone, that appears as a disturbance in any channel. Ghost roles often appear as gossip about someone not present. A ghost role acts as a dream messenger carrying messages for the health and well being of individuals and groups.

healing ~ To make sound, or whole; to relieve, restore or overcome; to "cure" according to the allopathic medical paradigm. "Healing" in this conventional sense, from a process perspective is a frequent side effect of "following" the deep natural flow of your life process with awareness. Healing from a purely process perspective is, in fact, an extraneous concept in the sense that life itself provides the "disturbances" to your health, and you can't be "cured" of your life. Following the deep sentient essence of your disturbances can, however, unleash inherent energies, and make them available for life and growth and living and healing.

immune system ~ The body's sophisticated security system that protects it from foreign substances and pathanogenic organisms and invaders, with both innate instant immune responses, and formulated instructive responses, both of which function below the level of consciousness.

levels of reality ~ Process Work concept of reality as consisting of three equaly important, and essential levels: consensus reality, dreamland, and essence.

metaskills ~ The deep feeling attitudes and beliefs about life and creativity, underlying and influenicng a facilitator's skill set and presence.

paradigm ~ A widely accepted belief or concept. An outstandingly clear or typical example, archetype or model. A philosophical and theorectical framework of a scientific school or discipline or worldview.

pathology ~ Scientific study of the causes of disease.

primary process ~ The primary identity of the person that is closest to their awareness at any given point in time. Also applies to relationships, families, groups, organizations, and communities.

process ~ The flow of nature.

Process Work/Process Oriented Psychology ~ A new awareness-based paradigm developed by Drs. Arnold and Amy Mindell and their associates, that facilitates individuals, families, organizations, groups, commuities, and eco-systems by following the flow of the deeper creative nature of the processes present in the moment. Process Work is science and art, linear and nonlinear, cognitive and experiential. Growing from its many ancient roots, Process Work is a new paradigm in relationship to our times.

secondary process ~ Aspects of ourselves that are furthest from our awareness, that we prefer not to identify with, that are "the other", and "not me"; but are disturbances to our normal identity/primary process, seeking our attention never-the-less. Also applies to relationships, families, groups, organizations, and communities.

sentient awareness ~ A deep inner expereince, usually an altered state of consciousness, in which language or any one communication channel is insuffient to describe the expereince. A deep awareness of the messages, the dreaming, and the meaning contained in sentient perceptions and sentient communications. Sentient awareness affects both the macro and the cellular realms of the body. Sentient awareness does not replace medical care. Sentient awareness is an act of reverence that compliments and enhances the whole person; mind, body, and spirit.

sentient communication skills ~ Finely tuned observation and facilitation skills. Sentient refers to very subtle communication signals, or portals/openings, that often lead to deeper communication, relationship, and healing by following their natural flow. Sentient signals that are noticed, supported, and nurtured respond with feedback. Following the flow of positive feedback unleashes inherent energies, and makes them available for life and living and healing.

Sentient Care™ ~ The name of Tom Richards' practice and his style of sentient awareness facilitation.

sentient essence ~ The spiritual dimension in which the world is not experienced as dualistic or marginalized or diversified or entangled, but as unified wholeness and coherence through sentient awareness. The closest human expereicnce to the infinite and the eternal.

sentient perceptions ~ Subtle, minute, flickerings and tendencies of body feelings, movements, sounds, and images. The realm of sentient or subtle tendencies that occur just before they can be identified with a communication channel. For example, noticing the impulse to rasie your arm before you rasie it, or were even aware you were going to raise it.

state ~ A process that is standing still, a static condition. For example, our "normal" state of consciousness in which we feel comfortable and realativley in control is, in fact, a habitual state of consciouness that is relatively static.

supervision ~ A confidential consultation for the facilitator about the patient and the process, with a respected colleague or mentor, to gain a new perspective and grounding before, during, or after a facilitation experience.

symptoms ~ Disturbances to our identity, of two types: accute and chronic. Accute symptoms are of short duration and typically severe. Many symptoms of short duration are not severe, but are never-the-less disturbing and early warnings, and are also referred to as acute. Chronic symptoms are symptoms that are continuous or recurr routinely or periodically revisit us over the long term.

wisdom of the process ~ The numinous flow of nature in which we are immersed, and with which we each co-create our life path.

Index

Publications   *Available at*: www.lulu.com/sentientcare
An Alzheimer's Surprise Party, Volume One, by Tom Richards and Stan Tomandl proposes that people with Alzheimer's dementia are not spiraling downward into "mindless pathology" as commonly believed, but are human beings in deep inner states of consciousness, parallel realities, that may reveal important and meaningful experiences for themselves, their families, and for society, even in their most extreme remote states, including coma.

An Alzheimer's Surprise Party Prequel, Volume Two, by Tom Richards and Stan Tomandl presents a profound example of an Alzheimer's patient working on himself psychologically, and communicating in meaningful ways with family, friends, and caregivers demonstrating personal growth and spiritual healing for patient, family, and caregivers.

Eldership: A Celebration by Tom Richards demonstrates the use of sentient caring skills to bring the treasures of eldership out from our seniors, many of whom are "stuck" in states of confusion, memory loss, high drama, delusion, delirium, dementia, Alzheimer's, end of life processes, and even coma; often lacking the awareness and communication skills needed to fluidly bring out their wisdom, experience, and treasure.

Eldership and Advocacy by Tom Richards explores personal barriers to social action and spiritual healing in palliative care situations, chronicling facilitation experiences, dilemmas, ghost roles, breakthroughs, and practical recommendations.

About the author

Tom Richards has evolved from a corporate executive with thiry-two years of experience in the health care industry, to a professional educator, researcher, and sentient care facilitator working with individual health issues, patient advocacy, and people in deep inner states of consciousness.

Tom is the author of *Eldership: A Celebration,* and *Eldership and Advocacy,* and co-author of *An Alzheimer's Surprise Party,* and *An Alzheimer's Surprise Party Prequel: Unveiling the Mystery, Inner Experience, and Gifts of Dementia.* All are timely and essential training tools to help transform the way people interpret and respond to patients with Alzheimer's, other dementias, and altered states of consciousness.

He earned a BS in electrical engineering from Cornell University, an MBA from the University of Chicago, and a Process Work Certificate from the Process Work Institute Graduate School.

Tom has studied, researched, and applied Process Work for over twenty years. He works with individuals experiencing body symptoms, injuries, illness, surgery; and memory loss ranging from mild cognitive impairment to Alzheimer's and other dementias, end of life delirium, and coma. Tom uses his awareness to follow people's processes to their deepest sentient levels, promoting healing; and encouraging eldership, growth, and beauty to blossom, at seemingly impossible times.

Contact information

Tom Richards

Sentient Care ™

1218 Roosevelt Avenue

Glenview, Illinois 60025 USA

Email: sentientcare@aol.com

Website: www.sentientcare.com

Publications: www.lulu.com/sentientcare

Marissa

Photograph by Lindsay Schlesser

www.ingramcontent.com/pod-product-compliance
Lightning Source LLC
LaVergne TN
LVHW091207080426
835509LV00006B/873